S0-CTR-534

When You Can Live Twice as Long, What Will You Do

When You Can Live Twice as Long, What Will You Do

and 99 Other Questions You May Have to Answer . . . Sooner Than You Think

Charles Platt

William Morrow
and Company, Inc.
New York

Library of Congress Cataloging-in-Publication Data

Platt, Charles.
When you can live twice as long, what will you do / Charles Platt.
p. cm.
ISBN 0-688-08686-1
1. Technology—Miscellanea. I. Title.
T49.5.P585 1989
303.48'3—dc20 89-31223
CIP

Printed in the United States of America

First Edition

1 2 3 4 5 6 7 8 9 10

BOOK DESIGN BY WILLIAM MCCARTHY

Acknowledgments

Most of the ideas in this book are derived from my own imagination and from small, miscellaneous references in science magazines and other media.

In a few cases, however, I have borrowed concepts directly from sources to which proper credit should be given.

Question 7. This is not speculation; it is fact. If you are interested in achieving immortality by having yourself frozen, contact the Alcor Life Extension Foundation at 12327 Doherty Street, Riverside, CA 92503.

Question 9. The subject of teleportation has been explored in many science-fiction novels, but in none better than *The Stars My Destination* by Alfred Bester.

Question 10. A thorough discussion of genetic markers is available in the excellent book *Genetic Prophecy* by Richard Hutton.

Question 23. The low-cost space drive exists as Project Orion. Its side effects would be exactly as I describe them, which is one reason its inventor, physicist Freeman Dyson, decided that it was socially unacceptable. The current treaty banning nuclear explosions in space also rules it out. Nevertheless, it would open up the solar system at considerably less social cost than we have paid for opening up the United States via the internal combustion engine.

Question 24. In the fall of 1988, West Los Angeles was sprayed with Malathion exactly as described in this question.

Question 31. As originally conceived, the concept

of SDI did entail sharing U.S. defense technology with the Soviet Union in the interests of maintaining a balance of power. Ronald Reagan referred to this concept, somewhat ineptly, in a presidential debate with Walter Mondale in 1984.

Question 41. The concept of homesteading asteroids is challengingly developed in Freeman Dyson's excellent book *Disturbing the Universe.*

Question 53. At the time of writing, it looks as if the FCC may deprive Americans of superior TV picture quality (already available in Japan) because of the lack of available broadcast frequencies—even though the new picture format could easily be transmitted over existing cable networks.

Question 56. These questions about drug use were inspired by the platform of the U.S. Libertarian Party.

Question 57. The UN treaty establishing common ownership of all extraterrestrial territory was proposed by the Soviet Union and will be a major disincentive to the development of space resources by private enterprise.

Question 58. This question is dramatized in Victor Koman's novel *Solomon's Knife* (Franklin Watts, 1989).

Question 61. This concept was first suggested in a science-fiction story by Damon Knight.

Question 70. I thank author John Shirley for suggesting this concept.

Question 71. Robot sentries of this type already exist.

Question 74. This concept was explored by Joe Haldeman in his novel *Tool of the Trade* (Morrow, 1987).

Question 83. This question was raised in *A Clockwork Orange* by Anthony Burgess.

Question 93. The Libertarian Party (which inspired this question) believes that armed free citizens are potentially less dangerous than armed governments.

Question 96. This concept was suggested by Mike Gunderloy in *Edge Detector* magazine.

Question 99. In actual tests of programs designed to simulate the comments of a psychotherapist, people were surprisingly willing to confess deeply personal problems even though they knew they were communicating with a computer.

My thanks to Merrilee Heifetz for inspiring this book; to Roberta Lannes for her comments and suggestions; and to Bob Wallace of Quicksoft for PC-Write, an excellent word-processing program with features that were of special help in writing and assembling the questions.

Contents

Introduction

Modern science is merely an accumulation of knowledge—a series of facts about ourselves and our surroundings. In previous centuries, these facts had relatively little impact on everyday life. The discoveries of Galileo, for instance, didn't much matter to an Italian peasant or a British nobleman. But as the store of scientific knowledge grew larger, engineers started exploring its practical applications, businesspeople started making money out of it, and major social upheavals occurred as a result.

Today, everyone knows that developments in electronics and biology may revolutionize our lives in the near future. The potential benefits are immense; but at the same time, we are becoming more and more aware of "the dark side of the force." For example:

- Modern medicine lengthens life. But when someone who is brain-dead is maintained on life-support machinery, we have to decide when it may be right to end life instead of prolonging it.
- Modern chemistry has brought us durable, affordable construction materials, fabrics, and furnishings. But each time we enhance our lives with these products, we subsidize factories that may be poisoning our environment with toxic waste.
- Modern genetics has revolutionized agricul-

ture, feeding millions who would otherwise starve. But if we experiment too freely with genetic modification, we risk releasing a life form that may start a plague.

These are the kinds of questions raised by technology. They are the side effects of science—ethical issues that previous generations never had to deal with.

Some of us find questions like these so unpleasant that we can't help wishing some kinds of technology didn't exist. It's tempting to argue, for example, that nuclear power is so potentially dangerous it should be banned.

But if we eliminate the side effects of science (such as the accidental release of atomic radiation), we will also lose its benefits (such as nuclear medicine). And in any case, history teaches us that science can never be successfully bottled up. Once a discovery has been made, someone, somewhere, will apply it; and a society that makes use of technology is likely to be more powerful than a culture that rejects it.

Most of us are resigned to the fact that there is a price to pay for progress. At the same time, we rely on "experts" (such as scientists, politicians, and lawyers) to grapple with the issues. We tend to feel that moral dilemmas raised by genetic research, for example, are too awesomely complex for us to deal with. So, we simply hope for the best.

But if we opt out in this way, we forfeit our right to participate in developments that concern us all. Can we really trust other people to come up with answers to these kinds of questions on our behalf? Wouldn't it be better to try to understand what's happening so that we can protect our own interests?

This book contains one hundred questions. They range from dilemmas that concern us right now to hypothetical situations that may not occur for decades—if ever. Some questions have unpleasant implications that aren't much fun to contemplate. Others are exciting, opening up possibilities that would have been inconceivable a few decades ago. All of them provide stimulating exercise in thinking about the impact of science on society.

Science itself is neutral—merely an accumulation of knowledge. But knowledge is power, and the more widely we share this power, the safer it will be. By becoming technologically literate, we guarantee ourselves a better chance of living in a future where we are masters of technology, rather than victims of it.

This is the true promise of science: knowledge that can bring us fulfillment and freedom. May we gain the expertise to use it wisely.

When You Can Live Twice as Long, What Will You Do

1 Should Public Officials Be Intelligence-Tested?

A new intelligence test is developed. It is free of all bias toward different ethnic groups and age groups. It is an all-around accurate test of reasoning power, creative thought, memory, and understanding.

Should politicians be required to take this test and publish their scores?

Should teachers take it? Doctors? Police Officers?

Should *everyone* have to take it?

Including you?

2 Would You Torture an Animal to Save Human Lives?

You are a research worker developing a revolutionary synthetic skin graft. It will help people with severe burns to recover in half the time, with less pain and no scarring. For some accident victims it could make the difference between life and death.

Before your new product can be used in hospitals, government regulations require it to be tested on live animals. Large areas of the animals' skin must be burned off before the skin graft is applied. A local anesthetic can't be used, because it might interact with the graft, and that would invalidate the data.

Will you personally conduct the tests?

If not, will you ask a lab technician to do it for you?

Should we do things to animals that we wouldn't do to people?

Should we ask other people to do things that we wouldn't do ourselves?

3 Would You Want an Electronic Brain Implant?

In the next century, advances in neurosurgery and biochips make it easy to implant a tiny "math module" in your head, via a small incision that can be done in a doctor's office with a local anesthetic.

You will communicate with this module merely by thinking about it. With its help, you will never forget a phone number again, and you'll be able to do complicated calculations merely by looking at the figures and wanting to know the result.

Can you imagine other kinds of modules that might be useful?

For example, do you have trouble learning foreign languages, remembering names and faces, or finding your way around cities you have never visited before?

What will happen if a module turns out to have a glitch in its programming? Or if it becomes accidentally damaged when you're depending on it to help you?

Which is more resistant to damage, tampering, or failure: a computer or a human brain?

Do we rely too much already on gadgets that do things for us? What's the alternative?

4 Who Should Control the Weather?

Suppose it becomes cheap and easy to produce a rainfall at any time and any place we want it. As a result, farmers have bumper crops, food prices go down, and everyone benefits.

But there's a fixed amount of water vapor in the atmosphere. It you create more rain in one place, there'll be less somewhere else.

If your country suffers a drought after a neighboring country starts inducing extra rainfall, how can you prove that the other country is to blame? And how can you force it to stop tampering with the weather?

Could weather modification be used as a weapon? Should it be banned by international law? How could such a ban be enforced? What would the penalties be?

5 Do You Want to Be Exceptionally Beautiful?

In the future, advanced techniques using computer-controlled "laser scalpels" and synthetic grafts make beauty affordable for almost everyone. The operations are simple, painless, and permanent.

Are you completely happy with every aspect of your appearance?

Which facial feature would you secretly like to change?

Will you feel less satisfied with your appearance if your friends start improving theirs?

If you remake yourself to look "like a model," will you look less distinctive than you do now?

Will it bother you if you no longer have a family resemblance to your parents? Will you mind if your children show no family resemblance to you?

If almost everyone becomes exceptionally beautiful, beauty will no longer be exceptional. Will that make it seem ordinary and less important?

Can you imagine a time when plain people became so rare they begin to seem special?

6 How Can We Throw Things Away When There's Nowhere to Throw Them?

In some cities in Japan, household garbage has to be sorted into five different categories so that it can be more easily recycled.

In the United States and most European cities, sorting is not required, and recycling programs are voluntary.

Japan is a group of small, overcrowded islands. Will Western nations face similar problems as they become more overcrowded?

How much garbage does *your* home produce?

How should we organize and pay for recycling our resources?

To get people to separate paper, metals, glass, and plastics from the rest of their garbage, we can introduce:

a. Laws, with still penalties for lawbreakers
b. Incentives, paying people for sorted waste products
c. A voluntary plan, relying on goodwill

Which system do you prefer?

If your choice is (a), who will enforce the laws, and who will decide how much garbage is too much?

If your choice is (b), who pays the reward money?

If your choice is (c), what will you do about families who won't cooperate?

7 Do You Want to Freeze Yourself?

It is already possible to preserve human tissue almost indefinitely by freezing it to the temperature of liquid nitrogen. Some scientists now believe that with proper pre-freeze preparation, memories (as chemical states in the brain) may also be preserved.

Some companies currently offer to freeze you after you die. Suppose there is a 5 percent chance that two hundred years in the future, it will be cheap and easy to unfreeze you, cure the disease that caused your death, and grow a new body for you at the same time. Would you give $100 a month, every month from now until your death, as prepayment for this kind of deep-freeze operation?

What if there was a 90 percent chance of being revived in the future?

What if some family members or close friends pledged to do it too, so that you might all wake up together?

Would you pay $100 a month for a drug that gives you a good chance at immortality? If so, in what way is this different from being frozen and revived in the distant future?

8 Could You Live in a World Without Privacy?

In the next century, an amazing spacetime scanner is developed that enables you to select and view any location, anywhere in the world. The device is cheap and easy to manufacture.

How will this affect nations where people are not normally allowed to know what their governments are doing?

How will it affect international politics, the stock market, and the arms race?

When it becomes impossible for nations to keep secrets from each other, will we live in a more peaceful world?

When people can't keep anything secret from their neighbors, will society be less hypocritical and more tolerant?

How will *you* feel when you can never be sure of keeping anything private?

What is your most personally embarrassing secret?

Your most professionally damaging secret?

Suppose that in this future world, a second gadget is designed that will *protect* you from being scanned. But it will cost you a year's income. Do you think it's worth the money?

9 How Far Would You Turn Back Your Biological Clock?

When medical researchers finally unravel the aging process, they may be able to turn back a person's biological clock to any given age and fix it there forever.

Would you like to be younger than you are now? If so, how much younger?

Can you imagine how society would be affected if no one ever had to grow old? How would seniority be determined? Would anyone be allowed to retire? Would people become less willing to take risks?

Population growth would become uncontrollable if everyone was immortal and could still have children. If a law was passed prohibiting immortal people from reproducing themselves, would you give up your immortality in order to start a family?

Could laws of this kind ever be successfully enforced?

Should research into the aging process be banned, for fear of the chaos that would result if immortality ever became available?

Would a ban of this type be effective?

10 Do You Want to Know Your Ultimate Fate?

Already, scientists can use genetic markers (patterns in your genetic code) to predict whether you will suffer certain hereditary diseases. In the near future, it will be possible to tell how likely you are to have heart attacks or some kinds of cancer.

In the far future, it may be possible to predict, precisely, your life expectancy and cause of death (assuming you die naturally).

Do you want to know how and when you are most likely to die?

Will this information change your everyday behavior?

Is life more fun if it's unplanned, without too many worries about the future?

Is life more fulfilling if you have a chance to look ahead and plan for tomorrow?

How will society change when every child knows his or her probable life expectancy?

11 How Much Do You Trust Your Intuition?

Imagine a drug that doesn't make you any more intelligent—but gives you stronger intuition. It helps you to "weigh up" a situation instinctively, enables you to be a better judge of character, and makes it easier for you to know intuitively what's best for you.

Is intuition more useful than intelligence when you deal with people in everyday situations?

Suppose that the drug only works on some people, but when it does work, the effects are permanent. Will you try it with your spouse or lover if there is a fifty-fifty chance that one of you will become much more intuitive while the other remains the same?

What will everyday life be like, dealing with people whose intuition is much better than yours?

Can higher intelligence help you to compete with people who have greater intuition?

12 Will You Risk a Flight into Orbit?

A second-generation space shuttle is developed by a private company. The company offers seats to paying passengers if they agree to fly at their own risk.

Suppose the fare is equivalent to your yearly income, and there is a 1 percent risk that the shuttle will crash. Will you pay the money and take the risk, for the thrill of seeing Planet Earth from space?

How much safer does it need to be for you to give it a try? How much cheaper?

Should the company be prohibited from offering seats to the general public if there is a high risk factor?

How safe should a spaceplane be before it is permitted to carry paying passengers? As safe as a modern airplane? As safe as airplanes fifty years ago? Less safe?

If a company can take money from people without accepting any liability for their welfare, is it more likely to be reckless and negligent?

Were airlines more reckless in the 1930s than they are now?

If regulators had demanded that airplanes in those days should be totally safe, would air travel exist today?

13 Do We Want an Electronic Democracy?

In the near future, using error-checking systems and secure passwords to eliminate fraud, people will be able to cast votes from their own homes via long-distance telephone lines. We can have an "electronic democracy" in which all major decisions are made by the general public, not by politicians.

Will an electronic democracy be fairer and less corrupt than the system we have now?

Should *everyone* be able to vote in this way? Even very young or very old people? People who can't speak English?

How do we decide who should be allowed to vote?

Will the general public, as a whole, be well-informed enough to make sensible decisions affecting the nation?

If we trust a doctor's judgment more than our own in medical matters of life and death, shouldn't we trust a politician's judgment more than our own in matters affecting the future of the country?

14 Could You Turn into a Pleasure Addict?

A new battery-powered gadget stimulates the pleasure centers of the brain. You can wear it as a small metal disk that adheres to your forehead, and it costs only pennies to run. There are no electrodes, no side effects, no shocks.

People who use it are more cheerful and optimistic. This positive attitude helps them fulfill all their ambitions. If they stop using it, however, they become depressed and withdrawn.

Will you be tempted to try it?

If so, how much will you be willing to pay for it?

Should children be allowed to use it?

At what age?

15 Would You Shut Down a Nuclear Power Plant?

You are a journalist investigating the safety record of a nuclear power plant. You discover that three years ago, an accident released a small amount of radioactive gas. Operators at the plant concealed the evidence—till now.

An independent expert tells you that the radioactivity was enough to cause, on average, one additional case of cancer among the 500,000 people living nearby.

If you publish your discovery, community pressure may force the power plant to shut down pending a full investigation. An old coal-fired power station will be brought back into service instead. The air pollution that it creates may cause deaths from lung cancer and emphysema.

Will you print what you have found out?

Suppose you confront the safety officer at the plant with the facts you have uncovered. He promises to revise operating procedures so that the accident can never happen again. Will you agree to kill your story?

In the past fifty years, how many people have died from side effects of nuclear power?

Why should the tiny chance of a nuclear catastrophe seem so much worse than the certainty of death and disability caused by air pollution from conventional power plants?

16 Will the Ultimate Lie Detector Be a Bane or a Boon?

Twenty years in the future, scientists perfect a "truth analyzer" that uses a combination of audible and visible cues to detect whether a person is lying. It is more than 99 percent accurate, and you can use it simply by pointing it in the direction of the person who is speaking to you. It can even be used on someone who is appearing on television.

How will this affect the business world? Government? Law enforcement? Husbands and wives? Parents and children?

How will you feel if the device is very expensive, so that government and big business can afford it, but you can't?

How will you feel if it's very cheap, so that every home can have one?

Will you want to use it on your friends, and people close to you?

Will this gadget make the world a better place to live in, once everyone gets used to it?

Do "little while lies" help to make life easier?

Even when people tell them to you?

17 Which Is More Disposable: Paper or Plastic?

You go to your local grocery store. The man behind the counter is willing to pack your purchases either in a paper bag or a plastic bag.

Paper is a renewable resource, and is biodegradable. Paper companies are replanting forests as fast as they chop them down. But our appetite for paper products causes temporary damage to the environment and kills small animals when their natural habitat is destroyed.

Plastic is derived from petroleum reserves that took millions of years to form. Once we use it up, there is no way to replenish the supply. And plastic products do not decay harmlessly after we dispose of them.

Which will you choose?

How much does it really matter?

Is it everyone's responsibility to protect the environment, or should big business do most of the work?

18 How Do You Decide If a Life Is Worth Living?

You are a doctor treating a man with head injuries. He has been in a coma for two weeks. His wife tells you that her husband often said he would rather die than live without all his mental faculties. She shows you a "living will" that he signed, stating this.

In your judgment, there's a 50 percent chance that he will never come out of his coma.

If he does regain consciousness, there's a 50 percent chance he'll be brain-damaged and will need constant care.

Will you follow the wishes of his wife and turn off the life-support system?

If the patient's family are very poor, making it almost impossible for them to provide care (or very rich, so they can afford round-the-clock nursing), will this affect your decision?

What if the life-support equipment is needed by other patients who might have a better chance of survival?

Will you make your decision as a matter of principle, or does it depend on practical factors?

19 Will Global Communications Bring Global Peace?

In the near future, new communications satellites will make it possible for one country to broadcast television programs directly to people in many other countries. Will it be a good thing for citizens in the Soviet Union to receive TV transmissions from the West, including uncensored news about their own government?

Will it be a good thing when people in your country can tune in to foreign transmissions encouraging people to question and overthrow your political system?

Do you generally believe in freedom of the press?

Should freedom of the press also apply to television?

Does that mean any adult should be free to watch any kind of TV program, from revolutionary politics to pornography?

If not, where do you draw the line?

Can a "free society" be totally free?

If you would want some things banned from TV, is it because they might have bad effects on other people, or because they might have bad effects on you?

The communications revolution began with the invention of the printing press. Have ordinary people benefited from it so far? Will they benefit more from it in the future?

20 Do You Want Unlimited Instant Travel?

Suppose that a combination of drugs and mental training enables people to develop their latent mental ability to move through space by willpower alone. This "teleportation" enables you to disappear from one location and reappear instantly someplace else.

Will you want to acquire this power?

Will you mind if other people acquire it too? Including people in other, poorer countries?

What will happen if people can relocate wherever they want, and national boundaries no longer exist?

But suppose you can only jump to a place after you observe it firsthand yourself. You must travel there by conventional means the first time around.

Will you want everyone's "teleporting power" to be limited in this way?

Will people in poorer countries agree with you?

21 Do You Want to Be Irresistible?

Scientists synthesize a real-life "love potion." It's a fragrance containing all the pheromones that cause sexual attraction.

One application of the product will last twenty-four hours, and it will cost you a week's wages. It should enable you to seduce anyone you want.

How much of your savings will you spend on this love drug?

If you use it, and it works, and then you stop using it, and your new lover loses interest in you, what will you do?

How will you feel if you find out that someone else has secretly used the product to seduce *you*?

22 Would You Try Out Your Own Wonder Drug?

You are a researcher studying a deadly virus that is sweeping across the country and causing thousands of deaths each month. You have found a vaccine that could stop the epidemic almost immediately, but you're afraid that in some people, the drug could cause the disease instead of preventing it.

An additional year of research should settle this uncertainty. But during that time, the virus will claim many more victims.

You believe you can prove the product's safety if you can test it on ten healthy, normal volunteers. Your assistant suggests that you and your staff should be the ones to try it.

Will you try out your vaccine on yourself, to save the lives of others?

Will you pay someone else to be the guinea pig? How much?

23 What Price Space Industrialization?

You are an engineer at a company developing new forms of rocket propulsion. You come up with an idea that will cut the cost of space travel so dramatically it will open up the mineral resources of the whole solar system within less than a decade.

There is a snag, however. There's no way to prevent your rocket engine from creating radioactive fallout. Each time it boosts a spaceship into orbit, the fallout will cause, on average, one additional human death from cancer somewhere in the world.

Will you ask for research funds anyway, downplaying the side effects?

Will you throw away your notes and start studying some other space drive instead?

How much global wealth has been created by gasoline engines in the past century? How many people have been killed by automobile pollution?

If we wait till our rare resources on earth have been used up, will we still be able to go into space to replace them?

24 Would You Spray a Neighborhood with Insecticide?

You are an official in the Department of Agriculture, trying to limit the area of your state that has been infested with fruit flies. If you fail in your task, the flies will cause huge amounts of damage to valuable crops. The economy will be affected, and everyone will suffer.

Studies show that a new population of flies is localized in a residential neighborhood of Los Angeles. Will you approve a plan to spray the entire neighborhood with Malathion? It is a pesticide that causes no known adverse health effects and has often been used over U.S. cities in higher doses that will be necessary here.

Bear in mind that if the flies are not eradicated right away they will multiply and spread to agricultural areas, where larger doses of pesticide will be required and fruit may need to be fumigated. Therefore, the total impact on the environment will be less if you act immediately.

How will you justify your action to the public?

If we want fruit that is reasonable in price and not infested with maggots, should we be willing to accept this kind of widespread application of pesticide?

Are there any other options that are as workable and effective?

25 Should We Eliminate Animal Species for Our Own Well-being?

You are a biologist employed by a utility company that wants to build a new dam. It will provide cheap hydroelectric power and will enable farmers to irrigate their fields. It will bring money to families in the area, most of whom are currently on welfare. Everyone wants the project. There is no local opposition.

Your job is to assess environmental impact. You discover that the river contains a rare subspecies of fish that will be rendered extinct by the dam. The fish is almost identical to other common species, so that if you say nothing about your discovery, no one else will know.

Will you sacrifice the fish for the welfare of the many people living nearby?

Should we stop any construction project if it threatens the survival of animal species?

Most of the Interstate Highway System was built before Congress passed laws to protect the environment. Under today's regulations, it could not be constructed. Does this mean that we now have a better relationship with technology?

26 Do You Want to Be Twice as Intelligent?

Drugs that are supposed to increase human intelligence are already available, though their exact effects are debatable. Imagine that in the near future, a new drug enhances your mental abilities by 50 percent.

What happens if a lot of people increase their intelligence in this way? Can you imagine the changes in government, industry, and everyday life?

MENSA is an organization for people who achieve high scores in intelligence tests. It so happens that very few leaders in politics or commerce are members of MENSA. Does this mean that you don't have to be intelligent to achieve success?

What do intelligence tests really measure?

Highly intelligent children are more likely to experience emotional and behavioral problems. Do you think it's true that "ignorance is bliss"?

Would you prefer a drug that promised to make you *less* intelligent (temporarily, at least)?

Do you ever drink alcohol?

If you had a few drinks before taking an intelligence test, how would the alcohol affect your score?

27 Would You Get Rich by Causing Chaos?

You are a physicist investigating subatomic particles. You discover a simple way to turn small amounts of matter into vast amounts of energy, more simply than ever before.

Using your technique, a small, cheap generator can produce almost unlimited power. Every consumer can own one. Oil and utility companies will be driven out of business. Individuals will control sufficient energy to transport them wherever they wish—maybe even into space. Life as we know it will be disrupted.

Will you suppress your invention for the sake of preserving law and order?

Will you turn it over to the government, because it entails a possible threat to national security?

Will you publish your discovery in a magazine, because you believe that information should be freely available, and any citizen should be allowed to build a power generator without government interference?

Will you patent your discovery, then obtain venture capital and start manufacturing the gadgets at a huge profit?

If some kinds of knowledge are too dangerous to be made public, whom do we trust to keep them secret?

28 How Much Would You Sacrifice to Save the Environment?

At some time in the future, it may be necessary for countries to consume less power so that less waste heat and carbon dioxide are released into the atmosphere. Otherwise, the greenhouse effect will melt the ice caps, raise ocean levels, and disrupt the climate.

You consume energy directly when you burn oil, gas, or gasoline. You use energy indirectly when you use electricity or purchase a new product—because a factory, somewhere, used energy in order to generate the electricity or manufacture the product.

Suppose the government asks everyone to consume half as much energy in order to save the planet. Will you cooperate willingly? If so, how much will your life change? Will you be less content than you are now?

Is an "energy tax" a better way to make people consume less energy? Even if it raises the cost of consumer goods?

If our nation introduces an energy tax, but other nations don't, can we still compete with them?

When we derive power from solar energy, we create no waste heat or carbon dioxide. There are no damaging effects. How can we encourage people to use solar power instead of burning oil and gas?

29 How "Perfect" Do We Want Babies to Be?

You and your spouse have decided to start a family. A pregnancy test has come back positive.

But your doctor discovers that you both possess a gene which creates a spinal deformity when it is transmitted to the fetus. In 90 percent of the cases, this deformity can be corrected by an operation after the baby is born. The remaining 10 percent are stillborn or severely crippled.

There is a 50 percent chance that the fetus has this defect. Imagine you are the mother-to-be:

Will you take a test that can tell you, for sure, whether your baby has the defect?

If the baby does have the defect, will you seek an abortion?

Modern medical technology is telling us more and more about the health of unborn babies. Will you want this information, or will you prefer not to know, and let nature take its course?

If you don't like the way nature works out, are you ready to live with the consequences?

30 Should "Dangerous" Knowledge Be Suppressed?

You are an American physicist attending an international conference. One evening, while talking with a Soviet scientist, you get an idea for a new high-powered laser. After the conference, you realize that the laser could be a devastating weapon. And the Soviet physicist has probably reached the same conclusion.

If your weapon is built, it will destabilize the balance of power and lead to a new arms race. Will you try to contact your Soviet counterpart discreetly to suggest that for the sake of world peace, neither of you should pursue this research?

Suppose he is the one who discreetly contacts you. Will you believe he is sincere, or will you assume that the U.S.S.R. will go ahead secretly and develop the weapon for its own advantage?

If your laser also has important peaceful uses—for example, in medicine—will this affect your decision?

31 Should America Share "Star Wars" Defense with Other Nations?

Suppose that American research finally results in a workable strategic defense system. New satellites can shoot down intercontinental missiles as soon as they are launched.

If the United States installs this defense system, the Soviet Union's missiles will become useless. Some strategists worry that the Russians may act rashly if their nuclear threat is going to be rendered worthless in the near future.

Should the United States refrain from deploying its new defense system, to preserve the balance of power? If so, can the United States trust the U.S.S.R. not to deploy a defense system of its own?

One advantage of a space-based defense system is that it can eliminate the danger of nuclear missiles that are launched by accident. Another advantage is that it can defend against missiles launched by other countries, so that nuclear proliferation is no longer a source of concern.

Ideally, the United States and the U.S.S.R. should both deploy strategic defense systems simultaneously. Then the balance of power would be preserved, and the nuclear threat would be eliminated at a single stroke.

Does this mean the U.S. should give its defense technology to the Russians for the sake of world peace?

32 What Would You Pay for a Good-Health Guarantee?

Imagine a new drug that acts as a barrier to *all* infectious diseases, from the common cold to AIDS. It's available for cash only. Medical insurance doesn't cover it.

How much will you pay for this kind of protection?

Ten percent of your annual income?

Twenty percent?

Fifty percent?

If supplies are scarce, should they be reserved for people who need them most?

How would this need be measured?

Who should have the power to make this allocation?

33 Would You Like Never to Be Sad Again?

Imagine an advanced technique derived from acupuncture, capable of controlling all human feelings by intercepting nerve messages to certain parts of the brain. Severe pain can be reduced to the point where it becomes easily bearable. Deep depression can be alleviated. Sadness can be erased.

But other feelings will also be inhibited. Happiness and pleasure will be reduced to a vague sense of complacency.

Do you want your pleasure/pain responses "evened out" in this way? Will it be easier to function on a daily basis if you always feel emotionally stable and nothing worries you very much? Will you sacrifice moments of great pleasure, just to know you will never feel unhappy again?

Do you envy people who seem naturally calm and untroubled—or do you want to feel things more vividly than you do now?

Do you believe authorities who claim that you need to feel very bad once in a while in order to experience intense happiness at other times?

34 What Do We Do When the Oil Runs Out?

When oil supplies begin to dry up in the middle of the next century, there's an urgent need for a new source of fuel. The cheapest alternative is "gasohol" made from fermented corn.

Because of population growth and changes in the global climate, many countries are experiencing severe food shortages. Exports of grain from the United States are crucial to prevent mass famine.

But farmers can make more money if they grow corn for gasohol. And the United States needs cheap fuel in order to maintain its economy.

Should farmers be paid by the government to grow food for export instead of corn for gasohol?

Should the United States reduce its consumption of fuel so that people in poorer countries can avoid starvation?

Is there any way to have both food *and* fuel?

Would it be wise for the government to support research into alternate energy sources now, to avert the potential crisis in the future?

Why are long-range plans of this kind seldom proposed?

35 Should Food Be Irradiated?

When food is exposed to low levels of atomic radiation, the radiation kills bacteria, and the food stays fresh longer. Consequently, less refrigeration is required and there is less risk of food poisoning.

The process seems absolutely safe. The food itself does not become radioactive. It remains just as nutritious as before, and its taste is not altered in any way.

Irradiated food can be cheaper and safer than ordinary food. But people are nervous about nuclear power. If manufacturers have to reveal on the label that packaged food has been irradiated, consumers probably won't buy it.

Would poor people benefit from cheaper food that didn't spoil so quickly and presented fewer health risks?

Should the government permit manufacturers to irradiate food without notifying the public, because the health and cost benefits are great, and opposition is merely based on superstitious fantasies about nuclear power?

Should people generally be given all the facts so that they are free to make up their own minds—even if they don't reach a rational conclusion?

Would you trust the judgment of a government regulator more than your own?

36 How Should Science Be Funded?

You are a politician drawing up a bill to allocate money for science.

If you spend it on research, you can make sure it goes to projects that tackle practical, short-term needs.

If you spend it on science education, you won't be able to control how the money is spent, and there will be no short-term benefits. But the long-term benefits will be greater.

You will be up for reelection four years from now. In the last election, you won by only a handful of votes. You need to show people that you have done something important for them during your term in office.

Will you be tempted to spend money on research because it produces a quick payoff that everyone will understand?

Suppose politicians such as yourself are not allowed to be reelected, and can serve only one term. Or suppose you are a government official in a country where there are no elections, and you serve for life. How would these factors affect the way you spend public money?

37 Do We Really Want 3-D Video?

A futuristic camera can tape your image holographically, in three dimensions, and play back a life-size replica of you accurate in every detail.

How much will you be willing to spend on this gadget? What will you use it for?

Suppose you want to make a holo-tape of yourself that people can look at after you die. What will you say to your descendants? Will you want to seem dignified or informal? Will you dress for the occasion or present yourself exactly as you are in everyday life?

If you could see a holo-movie of relatives who died before you were born, what would you hope to find out about them?

Would you want to see movies in genuine holographic 3-D, so that all the action seemed to happen around you? Would you want 3-D television? Newscasts? Game shows?

To what extent does the nature of a communications medium determine what people use it for?

38 Who Would You Swap Lives with for a Day?

In the next century, a communications network links almost everyone on the globe. Picture-phone lines have been installed in every country.

The service is so cheap, you can spend the whole day viewing and talking to a person on the other side of the world at negligible expense.

Suppose you can spend a day linked with one of the following: an Australian sheep farmer, a French novelist, a Soviet doctor, a Japanese fisherman, a Ugandan junior executive, or an East German police officer.

Whom do you choose (assuming they are all able to speak English)?

Will you learn more from someone in a less developed country, because of the contrast between it and your own?

What will the other person learn from you?

Will such a system encourage international understanding, or will it emphasize national incompatibilities?

Have global communications in the 1970s and 1980s helped to encourage international understanding? In what way?

39 Who Should Protect the Environment?

You are a farmer growing oranges in California. Your son is training to be a biochemist. He has synthesized a homemade insecticide that he claims will eliminate pest damage and increase the yield of your orchards by 30 percent.

Your son tells you he is convinced there will be no harmful effects on the environment. But he lacks the resources to perform the exhaustive tests that are necessary to get EPA certification. It's against the law to use any pesticide that has not been approved by the EPA. You must use the product illegally, or not at all.

Will you try the pesticide, knowing that there is virtually no chance of your getting caught?

Will you try it if you think there is any risk of damaging the land from which you earn all your income?

Do you believe that if the government doesn't protect the environment, no one else will?

40 Can We Sacrifice a Few Lives to Save Millions?

You are a senior administrator at a drug company. A project that you personally directed has produced a new birth-control pill. One dose will protect a woman for five years. In Third World countries where overpopulation is a leading cause of famine, this will save millions of lives.

The drug is now being marketed, and you have received a bonus and a promotion. But one of your researchers brings you new test results showing that in one case in a million, the drug causes cancer. It happens so rarely, no one is likely to discover the link.

Will you tell your researcher to keep quiet?

Can a wonder drug that causes very rare fatal side effects ever be acceptable, if it also saves lives?

Is any drug 100 percent safe?

If consumers are made fully aware of the risks, should they be permitted to use the drug if they still want to?

Should *all* drugs be freely available on this basis?

41 Would You Be a Homesteader in Space?

Asteroids are enormous rocks floating in space beyond the orbit of Mars. Preliminary studies indicate that some asteroids are rich in mineral resources. In the next century, it may be possible for pioneers to live entirely off these resources as "asteroid homesteaders" who never return to Earth.

Would anyone really want the risk, discomforts, and loneliness of life millions of miles from green grass and blue skies?

Five hundred years ago, long voyages by sailing ship were uncomfortable and dangerous, and many people feared it might be possible to sail off the edge of Earth. But explorers still set out into uncharted waters. Were they motivated by the promise of wealth? Discovery? Notoriety? Freedom from restrictions of everyday life?

Are there any places left on Earth where an explorer can find these things today?

Are people today less willing than they were five hundred years ago to face the excitement and dangers of the unknown?

42 Do You Want to Measure Your Pleasure?

As scientists learn more about the function of different areas of the brain, it becomes possible to measure activity in a person's pleasure centers using a simple clip-on sensor.

Movie makers can use the sensor to find out whether a test audience enjoyed what they saw on the screen. It can also provide a definite, objective way to measure the pleasure a person receives from a painting, a book, or even a love affair.

If the gadget is simple and cheap enough, it can be marketed as a novelty item—perhaps a sensor disk that runs through the spectrum, from blue to red, as the person's pleasure level increases.

Will you want your partner to wear it while being massaged, so you can learn how to provide the most pleasure? How about while kissing, or while making love?

How will you feel if you are the one wearing it?

When you can measure your pleasure, will this make it more difficult for you to enjoy yourself?

43 What Could Antigravity Be Used For?

Some theories predict the existence of antigravity—a negative gravitational force that pushes instead of pulls. An antigravity generator would nullify the attraction of Planet Earth.

If we had antigravity, anyone could own a hovercar. Highways would no longer be necessary.

What else could antigravity be used for?

If antigravity generators were installed at intervals throughout skyscrapers, would there be any limit to how tall the buildings could be?

Could most of the country's population be housed in huge floating cities, leaving the surface of Earth available for mining, agriculture, and parkland?

Antigravity would make space travel easy. As you move farther from Earth, its gravity gets weaker and easier to overcome. If you had an antigravity vehicle, is there any reason why you couldn't fly it to the moon or Mars?

44 Would You Want to Travel Beyond Tomorrow?

If someone offered you a round-trip ticket into the future on a time machine, would you accept?

Would you be more willing to make the trip if you had the option of staying in the future if you decided you liked it?

What if you *had* to stay in the future? (A drug to induce suspended animation would act as a one-way time machine of this type.) You might find a futuristic utopia—or the devastation of a nuclear war. Under what circumstances would you be willing to take the risk?

How many years ahead would you want to go?

If you could take four people with you, who would they be? What possessions would you take with you, if you were limited to one small suitcase?

45 Would You Peek at the Past?

Time travel into the past seems impossible, because it creates paradoxes. For instance, if you went back in time you could kill your own parents before you were born, then return to the present and find yourself still alive.

However, there is no theoretical reason why a time *viewing* machine should be impossible.

If you could look at a day in your own life, which one would you choose? A day from your childhood? A day in a love affair? A day in which you made an important good or bad decision?

If you could look back farther, which event in history would you select? A medieval battle? Settlers in the Wild West? The crucifixion of Christ? Dinosaurs? The creation of the universe?

If time viewing became available to everyone, what social changes would result? Would progress grind to a halt, as people became obsessed with the past? Or would we build a better, saner society, learning from lessons of history?

46 Would You Grow Your Own Clone?

A major problem in organ transplants is that your body tends to reject "foreign" human tissues. This is why close family members make ideal organ donors.

It may be possible to clone a human being—that is, grow an exact duplicate of a person from just a few cells. In that case, a wealthy man could grow a duplicate of himself, many years younger, and keep the clone drugged unconscious for spare parts only.

Is such a thing totally unthinkable?

Is it true that every time there has been a new scientific discovery, someone somewhere has used it for his own purposes, no matter how shocking?

Would large organizations have a use for clones?

For the ideal army, do we want independent individuals or obedient servants who all think the same way?

What kind of worker is most useful on a production line? In a typing pool? In a bureaucracy?

Might today's "unthinkable" fantasy become a reality tomorrow?

Can laws force people to use science only for ethical purposes?

47 Should We Build a Vacuum-Transit System?

Airplanes are able to fly faster at high altitudes because the air is thinner. Air resistance is a major factor restricting the speed of cars and trains on the ground.

Imagine a tunnel from which all the air has been extracted. A vehicle moving through this vacuum can theoretically travel at thousands of miles per hour. It can be faster and safer than an airplane. And it will use much less fuel.

But building tunnels to connect major cities is expensive. Even large corporations are reluctant to risk this kind of money.

Should the government finance projects of this kind, in the same way that government loans originally helped to construct railroads across America?

Should we let corporations take over the vacuum-tunnel system and run it at a profit, after it has been built? Or should the government run it, and return any operating surplus to the public?

If the government were running the airline industry and sharing its profits, would we all be better off?

48 Is the Space Program a Waste of Money?

If the money that we spent getting to the moon had been spent on research into medicine and agriculture, we could have saved millions of lives in less-developed countries.

Miniaturized computers and electronics were developed largely for use in space before they were widely applied elsewhere. Can we measure their benefits, now and in the future? Can we weigh these benefits against human lives?

Satellites have enabled better resource management and weather forecasts, helping farmers to protect their crops. Should we concentrate more on these practical uses for space, less on flashy "stunts" like the mission to the moon?

When the first satellite was put into orbit, did it seem like a "stunt"?

Photographs of Earth taken during moon missions made many people realize for the first time just how small our planet is. These pictures helped to create the ecology movement. How important has that been for our long-term survival?

Can we tell, in advance, which kinds of space exploration will produce substantial benefits?

49 Should We Allow Enhanced Athletes?

Professional athletes are prohibited from using drugs such as steroids and hormones. But suppose that some time in the future, a simple gene splice enables parents to rear unusually strong children. Should this be illegal? How could such a law be enforced?

Some races are genetically superior to others when it comes to physical strength and endurance. Black athletes, for instance, tend to do better than white athletes in many track events. Is this genetic inequality different in some fundamental way from a genetic inequality created by science?

If we want truly "natural" athletes, should we also restrict them from using special dietary supplements and scientifically designed exercise equipment?

50 Is Human Life More Important Than Religious Belief?

You are an obstetrician who has just delivered a baby for a woman whose religion prohibits birth control. She already has ten children. This child is her eleventh, and it is her fifth to be delivered by cesarean section. For various reasons, you firmly believe that if she gets pregnant again, another birth will kill her.

You suggest that she should have a hysterectomy so that she cannot conceive any more children. She refuses.

Suppose there is a new piece of equipment that can sterilize a woman by projecting a beam of high-energy particles. It looks just like an X-ray machine. You can use it on your patient, and she'll never know the difference.

Will you be tempted to do this? Remember, if you don't, she will very probably die as a result of her next pregnancy, leaving her existing children without a mother.

Should doctors ever do something for a patient's own good without admitting it?

Which is more important: a person's faith, or a person's continued healthy existence?

Is human life sacred? If so, are the life of a mother and the life of her child equally sacred?

51 Do We Want Dogs That Talk?

Experimenters have trained apes to communicate with people using a visual, symbolic language of one hundred words or more. Communication would be even easier if there were a bioelectronic implant that modified an animal's larynx so that it could imitate human speech.

Many dogs understand a variety of commands. Could we teach them more if they were able to have simple conversations with us?

Would talking pets be a fad that no one could resist? Or do people prefer pets that can't answer back?

A talking ape could be useful at work or at war, in dangerous situations where we prefer not to risk a human life. Is it moral to use animals in this way?

Would we have more trouble killing and eating animals if they could express themselves in a way we could understand? If cows could talk, for instance, would they still be routinely penned and slaughtered?

Is it true that the more a creature resembles us, the less we tend to think of it as a "mere animal"?

52 Would You Use the Ultimate Deodorant?

In the near future, a product is marketed that completely erases the smell from human sweat. It's cheap, and there are no side effects.

Will you want your spouse or lover to use this product?

Will you use it yourself?

Will you still want to use it if the effect is permanent?

Do you ever enjoy the smell of your own sweat? Of other people's sweat?

Will you feel differently about your body if it has no odor at all?

Studies show that people are sexually attracted to each other by body odors that we sense unconsciously. Will you be willing to give up this form of attraction for the pleasure of never having to worry about body odor again?

53 Will We Sacrifice Television Quantity for Picture Quality?

A new TV system is developed, offering wide-screen pictures in better color and detail than ever before. Movies at home can now look just as good as at a theater.

But there's a snag. To be transmitted by a TV station, the detail in the picture requires the equivalent of two channels instead of one. Since there are no extra channels available, some stations will have to give up their slots on the dial if this system is going to become a reality.

How can this be arranged?

Will some TV stations be willing to merge with each other and broadcast jointly?

Everyone complains that most TV programs are junk. Will we be better off if there are half as many programs, so that production companies can put more time, talent, and money into them?

Will this mean less programming for minority interests?

Which programs would you be willing to sacrifice, to receive better picture quality?

Does technology already have some effect on the artistic quality of mass entertainment? Will it have more influence in the future?

54 Would You Save Yourself Before the Rest of Humanity?

You are an astronomer making routine observations of distant stars. You discover something odd—a tiny point of light that is separate from the stellar background.

You check your figures and reach a shocking conclusion. You have discovered a comet that will collide with Earth six months from now, causing massive earthquakes, tidal waves, and climatic changes.

The comet is currently so far away that it's very hard to see, even with a telescope. Chances are no one else will notice it for two or three months.

If you reveal your discovery it may cause chaos as millions of people try to relocate in areas safe from quakes and flooding.

If you tell only a few close friends, you can quietly save yourselves by retreating to a safe area and fortifying your stronghold against the mass panic to come.

What will you do—save yourself, or tell the world?

If you choose to notify some friends, which ones will you tell, and which ones will you abandon?

55 Would You Want the Mental Power to Move Matter?

Some experiments suggest that our minds may have the power to exert a tiny amount of force on physical objects. By willpower, it may be possible, for example, to affect the fall of dice.

Suppose a research team discovers how to improve this telekinetic ability via a combination of chemicals and electronic stimulation. A very few people—two or three out of every million—can train themselves to lift objects weighing as much as an ounce by mental energy alone.

Will you want to be one of these special people?

Will it make any practical difference in your life?

Will you go to casinos and get rich at roulette, knowing that you'll win at the expense of other players?

Suppose you are a college student and you discover your mind power when you participate in a government-funded research program. Do you think you will be allowed to quit the program and return to a normal life if you want to?

If other people know you have this power, how will they treat you?

Who leads an easier life: a normal, everyday person, or someone who is very special and talented?

56 Should Recreational Drugs Be Legal?

A new vitamin pill goes on sale in health-food stores. It claims to be a food supplement, but people quickly learn that the main reason for taking it is that it gets you high. It has a similar effect to alcohol, but is non-habit-forming and nontoxic.

Should a drug of this kind be legal?

If it is used as a safer substitute for alcohol, won't it save lives?

Have drugs such as marijuana and cocaine always been illegal in this country? If they were legal once, why aren't they legal now?

For a drug to be sanctioned by the FDA, it must have therapeutic value. For a vitamin to be approved, it must function as a dietary supplement. Does alcohol fit into either of these categories? If not, why is it legal? Why are drugs such as caffeine and nicotine legal?

Should chemists be allowed to develop safer recreational drugs as substitutes for alcohol? If not, why not?

Whom are we protecting by antidrug laws: the person who wants to take the drugs, or everyone else?

57 Who Owns Outer Space?

An asteroid is discovered in an orbit that takes it close to Earth every ten years. An unmanned space probe discovers that the asteroid is about a hundred miles in diameter and is incredibly rich in metal ores.

A consortium of large corporations puts forward a plan to establish mines and factories on the asteroid. This way, industry can be relocated in space, removing major sources of pollution from Earth.

The corporations will develop the asteroid only if they can own it and make a profit. But under a United Nations treaty, no nation or private group is allowed to own extraterrestrial real estate. Asteroids, planets, and comets must remain freely available to all the people of Earth.

Was this United Nations treaty initially drafted by a socialist or a capitalist nation? Why did the United States sign it?

If North America had been declared public property before Columbus ever visited it, and all the world's nations had been forced by law to share the land among themselves, how do you think this would have worked out?

58 Should We Transplant Human Embryos?

A team of doctors invents a new technique enabling them to remove an embryo from one woman's womb and implant it in another's. This, they say, answers all arguments against abortion. In future, when a woman wants an abortion, the fetus does not need to die. It can be transferred to another woman who is infertile and wants to be pregnant.

Will right-to-life groups applaud this as a means of protecting the unborn—or will they condemn it as a violation of fetal rights and biblical commandments?

Will feminists be enthusiastic about a process that enables abortion without guilt—or will they be suspicious of a procedure that uses one woman as host to another woman's unwanted fetus?

Can you imagine any scientific breakthrough that might resolve the abortion debate once and for all?

Are there *any* issues on which scientists and religious fundamentalists can agree?

Must science always be in conflict with religious faith and dogma? If so, why?

59 What Will Happen If Life Is Found on Mars?

A new unmanned space probe lands on Mars. Using digging and sampling equipment, it discovers the remains of small creatures buried just below the surface.

Will this news cause excitement, panic, or indifference in the public as a whole?

How will the space program be affected? Will enthusiasm for space travel increase, or will people feel afraid of the unknown?

What are the implications for religious fundamentalists if it is proved that alien species once lived on another planet in our solar system?

Currently, there are plans for a joint American/Soviet manned mission to Mars. Do you think this is a waste of money? If traces of animal life had been found previously by an unmanned probe, would a manned mission seem more worthwhile?

What is the main reason for exploring other worlds: to search for signs of life, open up a new frontier for human settlement, develop new mineral resources, or merely investigate the unknown with an open mind?

60 What Skin Color Do You Want, If You Have a Choice?

A new chemical is discovered that lightens skin color. More than a mere dye, it penetrates the skin and permanently alters the pigment. Using this chemical in combination with minor plastic surgery, Asians, blacks, and other nonwhites can become indistinguishable from Caucasians, if they choose.

Will anyone want to use this product?

Will it be condemned as an insult to racial pride?

Will advertisements for it be banned in South Africa? In the United States? In Japan?

If differences in skin color are erased, will this do anything to reduce racial tensions? If not, why not?

Can science ever hope to remove some of the causes of conflict between national and racial groups?

Where prejudice exists, does rational argument ever make a difference?

61 How Would You Act If You Felt the Pain of Others?

Humans tend to feel empathy with other humans. For example, if we see someone who is very distressed, we tend to feel some distress ourselves.

Imagine a drug that heightens this effect. When you see someone in pain, you feel the same pain yourself. When you cause unhappiness in people by your words or deeds, it affects you as much as it does them.

Suppose a mad scientist puts this empathy drug in the nation's water supply. How will society change?

Will people still be able to work as police officers? Judges? Jurors? Doctors? Dentists?

Do people in these professions normally lack human empathy? Do they need to be cold-hearted in order to function?

If we were forced to care more deeply for the welfare of others, would the world be a better place to live in?

Would war become impossible?

Is it possible to go through life without causing pain to anyone?

If it's sometimes "cruel to be kind," is it sometimes kind to be cruel?

62 Should We Use Medicines Made from the Unborn?

You are a researcher studying tissue growth. You are especially interested in the way the body repairs itself. Why, you wonder, can someone grow a new fingernail, but not a new finger?

Unexpectedly, you discover a substance that induces the body to regrow parts that have been amputated. There's only one snag: the drug has to be extracted from human embryos.

Will you have the stomach to perform this procedure on aborted fetuses? Remember, the drug that you extract will enable crippled people to grow new limbs.

What if the aborted fetus must be at least twenty weeks old for the substance to be harvested? What if the fetus must be kept alive in a nutrient solution while the procedure is carried out?

If a fetus is not a human being, has no human rights, and cannot survive very long after being removed from a woman's womb, why should we feel squeamish about using it as a source of wonder drugs to cure people who are in desperate need of help?

63 Should We Answer a Message from the Stars?

For many years, scientists have scanned the sky for radio messages from alien civilizations. Suppose that sometime soon, the search finally pays off. Signals are received, recorded, and decoded as a symbolic language from an advanced race located just a few light-years away.

The signals promise that if we broadcast a reply, describing ourselves and our location in the galaxy, the aliens will trade information revealing important scientific secrets.

Should we send a message revealing ourselves to them?

Why would an advanced civilization bother to send signals out into space, trying to make contact with relatively primitive races such as ours?

Would a highly evolved species necessarily be more ethical than we are?

One of the early American space probes went out beyond our solar system carrying a metal plate embossed with pictures that showed our position in the galaxy. Do you think that sending this into the unknown was a good idea?

64 Will Literacy Become Obsolete?

In the next decade, speech recognition systems becomes widely available. As a result, computers can now understand us when we speak to them.

When you want to write a letter, you dictate it to your word processor and it prints the text onto paper for you. To take money out of the bank, you speak to the automatic teller machine. No one needs to type on a keyboard or write by hand anymore.

Does this mean that in the future, we no longer need to teach children how to write?

Electronic calculators have already made it unnecessary for people to practice their mental arithmetic. Consequently, many of us are less skilled at addition than we used to be. Is this a bad thing?

Why should we drill ourselves to perform mechanical tasks that a cheap machine can do more efficiently?

Some schools still teach Latin, even though no one speaks it. Will a few institutions continue to teach handwriting in the same way, after it is generally obsolete?

Would you want your children to learn this kind of "lost art"? If so, why?

65 Would You Share Another Person's Total Life Experience?

Future advances in brain research make it possible to record one person's brain activity—and play it back into another person's head. This way, you can experience in every detail the sights, sounds, tastes, smells, and emotions perceived by another human being.

Will you want to try this?

Will you want to play a tape made by an astronaut, a racecar driver, an undersea explorer, a movie star, or a surgeon, so that you can share his experiences?

Are there any people whose minds or experiences you will definitely *not* want to share?

Will you want to run a tape made by someone you know?

A close friend?

Someone of the opposite sex?

Will you allow a tape to be made from your own mind, for an hour? Who will you be willing to share it with?

66 Would You Commit Electronic Fraud?

You are the parent of an eighteen-year-old computer whiz kid. One evening you find that he's left his computer terminal on. You look at the screen full of figures and realize with a shock that your son has tapped into a network that transmits funds electronically between banks.

When your son returns, he admits that he has been skimming $1 from every transaction in the network, sending it to an account of his own. Because there are thousands of transactions every day, he has accumulated more than $1 million in his account. As yet, no one suspects anything.

What can you do that is morally right, short of turning your son over to the police?

Will you make your son close his account and give the money anonymously to a charity? (It is not possible to return the money to the people it was stolen from, because so many transactions were involved.)

Does electronic fraud somehow seem less serious than other kinds of crime?

Will you be tempted to tell him not to steal any more in the future if he'll share some of his $1 million with you?

67 Will You Drive a Solar-Powered Car?

New developments in solar cells make it feasible to build a solar-powered car. Its maximum speed is only thirty miles per hour, and its batteries give it a range of only twenty miles if the sun isn't shining. But it is completely pollution-free and does not deplete energy resources.

If everyone starts using these vehicles for short trips in urban areas, pollution in our cities will be cut in half, noise will be reduced, and oil consumption will be drastically curtailed. Everyone will benefit.

But how can people be persuaded to give up the speed, convenience, and range of a gasoline-powered car?

The government can subsidize solar-powered vehicles to make them so cheap that people will want to buy them. Should we increase everyone's taxes to pay for this subsidy?

A stiff tax can be imposed on gasoline-powered cars. Does this mean we should pay compensation to people who need them to travel long distances in rural areas?

Gasoline-powered cars can be banned from cities. Does this mean we should compensate people whose cars are no longer as useful as they used to be?

Is there any easy way to encourage people to make individual sacrifices for the benefit of the whole community?

68 How Much Do You Really Want to Eat?

A new kind of prepared food contains three times the calories of an ordinary meal, plus an enzyme that slows digestion to one-third its normal rate. This means you can eat one meal at the beginning of the day and feel full and satisfied all the way through to bedtime.

Will you value the extra time in your day, now that you need to eat only one meal instead of three?

Do you usually eat because you're hungry? Or because you relish the taste of food? Or because of the sensual experience of tasting, chewing, and swallowing?

Do you distrust high-tech innovations in your eating habits? If so, do you stay away from instant breakfasts, TV dinners, and low-calorie sodas?

Eating is a primitive, instinctual activity. Do we tend to be conservative and distrustful of science when it is applied to this kind of area?

Do you get more pleasure out of life when you follow your intuition and do what feels good, or when you decide what's best intellectually?

Is there any way to design a scientific diet aid that will really work?

69 Would You Like to Be Quicker-Witted Than Anyone Else?

You are a chemist experimenting with organic compounds. You come up with something that you believe will increase the speed of nerve impulses in the body. You try it on yourself—and it works!

Your product not only gives you super-fast reflexes. It also doubles the speed of your brain. You are not more intelligent than before, but you think twice as quickly.

Will you sell your invention to a drug company, and make a fortune?

Will you keep it to yourself, so that you can have an edge over everyone else?

Your extra speed may help you to be a champion racecar driver, a game-show winner, a successful executive, or a test pilot. In what other occupations will it give you an important advantage?

Is quick thinking more important than intelligence? Is it necessary in order to succeed in life?

If you think twice as fast as normal people, will you enjoy feeling superior, or will you feel impatient and isolated? Will you end up wanting to share your discovery with other people, even though you'll no longer be so special?

70 Would You Fight Drugs by Polluting the Water Supply?

You are a high-ranking law-enforcement official trying to stem the rising tide of drug use. You genuinely want to achieve a drug-free society; yet each time a dealer is arrested, others take his place.

Through unofficial channels you learn that a chemist has synthesized a harmless substance which causes nausea when it is used in combination with narcotics. If people consume a small amount of this substance each week, they'll feel horribly sick anytime they try to use illegal drugs.

You want to add this substance to the water supply in major cities. But your plan will be delayed for years by FDA certification and opposition from civil-liberties advocates.

Will you be tempted to have the substance added to reservoirs secretly?

Is it sometimes right to break one law in order to enforce another?

Do you believe that if the public knew about your plan, most people would approve of it?

In a democracy, are the rights of innocent, law-abiding people more important than the rights of drug users?

If so, does that mean you have an obligation to protect the majority using any means available?

71 Will You Want a Robot Watchdog?

A manufacturer of security devices introduces an expensive new gadget: a robot watchdog that patrols your property, challenges any intruder, and disables burglars with painful, piercing sound and stroboscopic lights.

The robot has no weapons. The worst it can do is scare people and inflict temporary pain on their ears and eyes.

Do you think homeowners are entitled to protect their property with this kind of device?

Is it much the same as a real watchdog?

If a friend pays an unexpected visit, ignores your warning signs, accidentally triggers the robot, panics, runs, falls, and breaks his ankle, whose fault is it?

Ideally, should the robot be more intelligent, so it can distinguish friend and foe?

Should it speak and understand English, and make simple decisions on its own initiative?

When a system is more complicated, are there always more ways in which it can go wrong?

If intelligent robots ever become common, how will we make them do what we want? How will we protect ourselves from their poor judgment and errors?

72 When Is It Right to Do Weapons Research?

You are a physicist looking for employment. Finally, you receive an offer from a defense contractor. The contractor wants you to design a new kind of nuclear device that creates more blast damage and more radioactivity than conventional bombs.

You know this device will never be used unless there's an all-out nuclear war. And if that happens, humanity will be wiped out anyway.

Will you take the job, because your family needs the money and you believe the nation needs an effective nuclear deterrent?

Will you take the job, but only until you can find nonmilitary work elsewhere?

Will you take the job, but deliberately introduce errors in your work to save your pacifist conscience?

If you are perfecting a deadly device that is unlikely ever to be used, are you putting society less at risk than if you do pure research that may have unpredictable military applications in the distant future?

Can a physicist ever be sure that his work won't be used in some way to kill people?

How much responsibility does a scientist have to do research that can't be used for military purposes?

73 What Would You Do If Armageddon Were Four Years Away?

A leading astrophysicist announces that the sun has become unstable. It will turn into a nova just four years from now. The atomic reactions are already taking place, and disaster is inevitable.

When a star goes nova, it explodes. All life on Earth will be incinerated. There can be no escape, even by hiding deep underground.

Assuming that other scientists agree with the prediction, can you imagine how you will react to the news?

How do you think other people will react?

Will religious people act differently than people who don't believe in God? Will people from other countries behave differently?

Will life go on much as before—for the time being at least—or will society fall apart as millions of people quit their jobs and abandon their homes?

Will social chaos wipe us out before the nova actually occurs, or will life degenerate into an orgy of indulgence, or will people face the end with calm dignity?

74 Would You Make People Do What You Want?

You are an amateur electronics enthusiast building a music synthesizer. By chance, you stumble on a combination of ultrasonic frequencies that has a numbing effect on the brain. Using trial and error, you find that when people hear the ultrasound it puts them into a brief, light trance. While they are in the trance they do whatever they are told.

You make a miniaturized, battery-powered version of your gadget and take it outside, wearing earplugs to protect yourself from its effects. Now you can go to a bank and compel the teller to give you as much money as you want. Or you can make a salesman give you a new car.

What else can you do with your gadget?

Will you be tempted to use it on your enemies?

Will you get a taste for power, and want more and more?

Will you use your gadget to make people fall in love with you?

Will you end up jaded, corrupt, but afraid to stop using the gadget in case you lose everything you've gained?

Do the side effects of an invention sometimes outweigh its benefits?

Looking ahead, will you be tempted to throw the thing away and forget all about it?

75 Do You Want to Sleep Half as Much?

Imagine a simple gadget that induces deep sleep and maintains you in a state of total relaxation. Because you are sleeping more efficiently, you feel rested and refreshed with half as much sleep as before. Suddenly, you can have extra waking hours in every day.

Will you be tempted to try this gadget?

Will it bother you if it prevents you from remembering any of your dreams?

If everyone starts using it, no one will need much sleep. How will this change social habits?

Will there be more stress in families as people have more time to get on one another's nerves?

Will people get bored with their extra leisure time?

Will employers start wanting people to work longer hours?

If you have an extra four hours in every day, how will you use them? Will you be happier and more fulfilled?

Will you quickly get used to the extra hours, and feel no better off than you were before?

76 Should We Use Intelligent Animals as Servants?

As we learn more about the genetic code, we find out how to manipulate it to breed animals with traits that we find useful. Cows can be made to give more milk. Chickens can be more resistant to disease. Horses can be stronger and faster.

Should it be illegal to "play God" and breed an animal to serve our purposes?

Cattle ranchers routinely mate animals that are most likely to breed valuable livestock. Is this kind of genetic selection wrong, too?

Suppose we design an entirely new breed of animal that stands on two legs, is four feet tall, has a friendly disposition, is strong but obedient, and eats almost anything. It understands English and can learn simple tasks.

Will you feel comfortable using one as a domestic servant?

Will you want to see these animals carrying out menial tasks such as street cleaning or farm work?

Will a semi-intelligent animal of this type have "human rights"? Will it have free will?

Is it wrong to create an animal that is truly happy when it works hard for its master?

Was it wrong for humans to tame wild dogs and horses?

77 How Much Will You Pay for Highway Safety?

In the next decade auto manufacturers start offering a radar-controlled collision-avoidance system as an optional extra. The system detects any object on the highway ahead and automatically applies the brakes if necessary.

Experts estimate that if every car is fitted with the system, it will save a thousand lives a year. Will you pay $2,000 to have it installed on your car?

Will you pay $500? $100?

How do you balance cost against safety? Can there be a dollar value on human life? Should we even consider cost, if lives can be saved?

If you install the collision-avoidance system, will it make you feel safer at high speeds? Will that tend to encourage you to drive a little bit faster?

If you rely on a system of this kind, do you become less conscientious and skillful as a driver? If the system fails unexpectedly, are you worse off than before?

If the system applies the brakes when there's no emergency, can it cause an accident instead of preventing one?

Should we continue developing systems of this type?

78 How Would You Feel If Science Discovered the Human Soul?

Suppose that ten years from now, scientists detect a shifting electrical field above a person's body immediately after death. Further tests prove that a kind of magnetic entity leaves the body when someone dies. Science, it seems, has discovered the soul.

Will this alter your view of the world?

Will you feel comfortable about this kind of research?

Suppose you are suffering from a fatal illness and feel you may be near death. Will you allow the monitoring system near your bed?

How will you feel if scientists use electromagnets to trap someone's "soul entity" so that they can study it?

Does the soul have human rights?

If the soul is immortal, why should we worry if scientists decide to study, capture, or interact with it?

79 Would You Rather Live in Another Universe?

According to the quantum theory, when we try to observe individual particles such as electrons, an "uncertainty principle" means we can never know exactly where a particle is or how it will react with a single other particle.

Some people believe this means there are many different results of every reaction at every instant, creating multiple realities that branch from our universe all the time.

Suppose someone invents a gadget that lets you hop to a parallel universe where history is slightly different.

Will you want to skip to a universe where Kennedy wasn't assassinated? Where America defended Pearl Harbor? Where the South won the Civil War? Where Greek civilization was never destroyed?

How do you think history will be different as a result?

Will you want to live in a universe where your own life is different? If you could change one event or decision in your past, what would it be?

Do you believe that there is really only one path that we are fated to follow? Or is our destiny under our own control at every moment?

80 Would You Design Your Own Child?

It should soon be possible for parents to choose whether to have a boy or a girl.

If genetic manipulation is allowed, you will also be able to control other aspects of your child. Will you want your son or daughter to be blond or brunette, blue-eyed or brown-eyed, short or tall?

Do you dislike the idea of meddling with the process of bringing new life into the world?

If so, does this mean you won't do *anything* to influence the development of a fetus?

Suppose you especially want a boy baby. Some evidence shows that you can marginally affect your chances of conceiving a boy or a girl by having the woman rest in one position or another after intercourse. Do you regard this as interfering with the natural process of conception?

The extent to which we let scientific knowledge intrude into the creation of life must be a matter for individual conscience. But are there any rational guidelines to help us decide these difficult questions? Is it purely a matter of instinct and intuition?

81 How Could We Survive Without Electricity?

As a result of a wandering stellar object that enters our solar system, Earth is subjected to very intense magnetic fields. The fields are so strong, they make it impossible for us to use any electrical appliances on Earth.

How many times in your daily life do you use electricity directly or indirectly?

How many of the items you buy, from clothes to food, would be difficult or impossible to manufacture without electricity?

What would happen to modern civilization if this disaster scenario came true, with very little warning, and electricity was no longer available to us?

How uncomfortable would life really be without electricity? If people from one hundred years ago could see the way we live today, would they agree that electricity is essential to civilized existence?

To what extent have electrical gadgets improved the quality of our lives?

82 Would You Travel to Another Star?

Fifty years from now, humanity builds its first interstellar spaceship. The vehicle will set out on a forty-year voyage to another star, which has its own family of planets. One of these planets, at least, may be habitable.

The ship will carry one hundred colonists for the new world, plus equipment to establish farming and industry. Suppose you are under thirty years old. Will you want to make the trip?

If you could be put in suspended animation so that you would emerge at the other end the same age as you are now, would this affect your decision?

Can you see any practical reason for colonizing the planets of other stars, bearing in mind that it must always take a good part of a lifetime to travel to and fro?

Does it concern you that if we wipe ourselves out in an atomic war, and there is no other outpost of humanity, the human race will become extinct?

To what extent do you feel humanity has a deep need to "be fruitful and multiply"? If the human race spreads out across the galaxy, will you see this as a curse or a blessing?

83 Can We Program People to Be Good Citizens?

Imagine a new corrective technique that can actually eliminate criminal tendencies. Using principles of behavioral psychology, volunteer convicts are subjected to a mixture of rewards and punishments designed to make them feel physically ill if they ever try to break the law again.

Is this kind of conditioning a better system than keeping people in jail, where they remain unproductive and must be supported with public money?

If someone stops committing crimes because he has been conditioned not to, has he really been cured of his criminal tendencies, or has he merely been brainwashed?

If a child is brought up by strict parents who punish him severely if he steals, isn't this a form of brainwashing?

If we allow a prison system to reprogram the minds of disobedient citizens, is this a first step toward establishing a police state?

Should a person always be free to choose between good or evil?

If he is deprived of this freedom of choice, is he no longer truly human?

84 How Would We React to the Totally Unknown?

Suppose that a large, mysterious object appears suddenly in the desert. It is a shiny metal sphere, fifty feet in diameter. Its origins are a total mystery. It may have come from outer space, or from another dimension.

The sphere is featureless. It seems to have no purpose. It does no harm to anyone. It does not respond to attempts to communicate with it, and scientific instruments can detect no sign of life inside it.

You are a scientist leading a team of investigators. What will you do? Will you try to cut a way into the sphere, X-ray it, destroy it, or merely watch and wait?

When we are confronted with something that seems frighteningly alien and inexplicable, what do our survival instincts tell us to do? Should we follow these instincts, or should we try to apply logical analysis?

Imagine an ant finding its path suddenly blocked by a discarded Styrofoam cup. Even if the ant is intelligent, can it hope to understand what the cup is for and where it came from? How should the ant deal with this invasion of its territory?

Would we be better than the ant at making sense of something that came from completely outside our world?

85 Will You Trust an Expert System More Than a Human Being?

An "expert system" is a computer that imitates the thought processes of an expert who uses specialized knowledge to solve a problem. Expert systems advise oil companies where to drill, based on complex geological data. They will soon assist medical diagnosis.

Suppose you are a doctor trying to find out what's wrong with a patient. You tell the expert system what the patient's symptoms are, and it suggests a likely cause based on standard medical knowledge.

You think the diagnosis seems very unlikely. Do you accept it anyway, knowing that an expert system takes more factors into account than you can?

Suppose that the drugs recommended by the system will cure the patient if the diagnosis is correct—but will cause severe complications otherwise. What do you do?

If you were the patient, would you rather trust the opinion of one man, or a computer program that incorporates the accumulated wisdom of hundreds of specialists?

Is an expert system likely to slip up on a very simple, obvious diagnosis?

Do human beings sometimes make foolish blunders because they overlook something that ought to have been obvious?

86 Should Children Have Artificial Playmates?

In the next decade, there are lifelike dolls that can speak a vocabulary of five hundred words. Given a few verbal cues, they can formulate simple sentences in reply.

Kids love the dolls because they're so much easier to deal with than flesh-and-blood playmates. A doll doesn't hit back, doesn't complain, doesn't get bored, and doesn't ever disobey.

Will we allow children to own this kind of toy, even though it encourages them to retreat from the real world?

Can a doll be programmed to simulate the active give-and-take of a real human relationship, or must it always be a passive partner, feeding its owner's ego?

Many adults enjoy playing with electronic gadgets. Why is this, exactly? Is it healthy, mature behavior?

Some computer users become obsessed with their high-tech toys. Does this matter, if it makes them happy and they earn a good living out of their obsession?

Which would you really rather live with: a quirky, irritating person, or an obedient adult-size doll that's always entertaining and obeys your every whim?

87 Would You Want a World Where There Was No Place to Hide?

Your body's genetic "signature" is unique. This means that your unique pattern of genes can be used for purposes of identification.

When everyone's genetic data are recorded at birth, police at the scene of a crime only need the tiniest clue to identify the felon. A single human hair is more than enough.

Will this make you feel more secure?

Or will you feel more secure if your genetic data are *not* on file with the authorities, because that way they won't know so much about you?

Imagine a monitoring system that continuously samples and analyzes the air in public places. It can identify each person who walks past, using the genetic chemistry of microscopic particles that we shed as we move. Large centralized computers continuously process these data, so that no one can hide from the police.

Is there any existing law to prevent authorities from installing this kind of surveillance system?

Would there be a public outcry against it, or would people welcome it as a way to eradicate urban crime quickly and permanently?

88 Would You Make Illicit Use of Information in a Database?

You are an ambitious young executive angling for a promotion. You have one powerful rival—a coworker whom you suspect of using unscrupulous tactics against you.

Your coworker often brings his computer from home to the office. One day he leaves it running while he's away from his desk, and you start playing with it out of idle curiosity. Unexpectedly, you get into a database on his hard disk and discover a lot of personal financial information, including payments to a psychiatrist and a treatment center for alcoholism.

Will you find a way to use this personal information to discredit your rival, knowing that he might do the same to you if he had the chance?

Will you confess to your rival but promise to keep quiet, and hope to earn his goodwill in return?

Will you simply say and do nothing about it?

If the FBI obtains confidential financial information about you and uses it against you, is this any different, in principle, from your action against your rival?

89 Do You Want to Be Weightless?

In the next century, you visit a space station in orbit around Earth. Up there, you experience zero gravity. You and all the objects around you are in "free fall."

Do you think this will feel pleasant?

Which daily activities will be easier, and which will be harder, when you don't have gravity pulling you downward?

Do you think you'll get accustomed to moving freely in any direction, so you'll feel clumsy and restricted when you come back to Earth?

On the moon, gravity is only one-sixth as strong as it is on Earth. Will you want the power and freedom that comes from weighing one-sixth as much as you do now?

Suppose you suffer from circulatory problems. You'll live much longer under reduced gravity. Will you be willing to abandon Earth and live in a lunar colony under an airtight dome, if it will add twenty years to your life?

Can you imagine a future time when people have adapted themselves so successfully to life in space that the high gravity of Earth makes it seem like a prison?

90 When You Can Live Twice as Long, What Will You Do?

A new drug is announced that enables you to live longer by slowing the deterioration of cells. Scientists estimate that average life expectancy will increase to 150 years as people grow old half as fast as before.

You are one of the lucky ones who can afford the new drug. Suddenly you have twice the life expectancy that you used to have (assuming you don't die in some form of accident).

How will this alter your way of life?

Will you be more likely to put things off, knowing you'll have more time to take care of them later?

Will you tend to hurry less, be less competitive and more relaxed?

Imagine seeing people around you who can't afford the drug, growing older twice as fast as you. Will this make you feel good or bad?

If people close to you refuse to use the drug, will you still go on taking it?

Will you be tempted to take it secretly?

91 Do You Want to Cheat Death Inside a Computer?

Your memories and your personality exist as a pattern of information inside your brain. Imagine that scientists are able to scan this information a bit at a time. This way they can copy it into a computer's memory.

When your consciousness lives on inside a computer, you become immortal (so long as no one switches the computer off). The computer supplies you with sights and sounds from the outside world, and it also creates internal simulations of reality for you. Within limits imposed by the size of the device, it gives you any experience you desire.

Do you like the idea of your consciousness existing as a pattern of electrons inside a machine?

In what way is this different from consciousness as a pattern of electrochemical states inside your brain?

Would you feel deprived if you could no longer experience feelings from your physical body? Even if the computer could simulate, accurately, any physical sensation that you wanted?

Do you believe the soul may be immortal?

How does this differ from immortality inside a computer?

92 What Can We Build with "Supersteel"?

Imagine a new material, one hundred times as strong as steel, but one hundred times as light. If it's cheap enough, what can it be used for?

Are the strength and weight the major limitations in the architecture of buildings? The size of airplanes? The length of bridges? The design of space vehicles?

Superstrong material could even be used to build a "beanstalk" into space. This would be a cable hanging down from a geosynchronous satellite, with its lowest point anchored on the surface of Earth. (To make it stable, the cable would also have to extend up beyond the satellite, terminating in a counterweight.)

If we built a beanstalk, we would no longer need rockets to escape Earth's gravity. We would simply ride an elevator up the cable. Studies have shown that such a device would be feasible, given some advances in materials technology.

Exactly what would happen if something went wrong and the "beanstalk" broke halfway up?

93 What Happens When Everyone Is Armed and Dangerous?

You are a physics instructor at a college. One of your students accidentally leaves his notebook behind after class. Leafing through it, you find plans for a project that he has been building in his own time.

It's a device that will emit a beam of intense radiation, designed to interrupt the functioning of a person's brain. A short burst will render someone unconscious; a long burst will be lethal.

Your genius student is creating the ultimate handgun. It will be quick and silent, and will leave no evidence of any kind.

Will you try to persuade him not to build it?

Will you try some other means of suppressing the invention? If so, what?

If everyone owns a weapon of this kind, able to inflict instant death without leaving a trace, will the world be very different from the way it is today?

In Switzerland, there are guns in every home. Is Switzerland considered a dangerous, violent country?

Our government controls the largest weapons. It also supervises military research. If science gave more power to individual citizens, would this create a healthier balance?

94 Will You Want the Ultimate Dieting Aid?

Imagine a little electronic gadget that suppresses a person's craving for food. You wear it like a hearing aid. Anytime you feel tempted to eat a snack, you reach up and switch on the gadget. It broadcasts a brief signal to your brain, suppressing your hunger.

If you need to diet and the gadget is cheap and has no side effects, can you see any reason why you won't use it?

Do you think you will have the willpower to reach up and switch it on?

Will there be moments when you won't want your appetite to be suppressed, because you'll want to satisfy it naturally?

In order to overcome a physical craving, do we first need a gadget that will increase our willpower?

95 What If UFOs Really Do Exist?

At first it sounds like an item out of the *National Enquirer*: an alien space vehicle has crashed in Wyoming.

A team of investigators rushes to the scene. They find that the aliens have erected an invisible force screen around their vehicle, preventing anyone from coming too close. Repairs seem to be under way. And within a matter of hours, the UFO takes off. It easily outruns Air Force jets and disappears into space.

Scientists now have proof that UFOs do exist. But no one knows anything more than that.

What effects do you think this news item will have, nationally and internationally?

Will people become nervous and fearful?

Will nations tend to cooperate more, to build strength against the unknown?

Will we want to pursue space exploration more actively, or will we be more likely to turn away from space, for fear of what we may find out there?

Which is the more powerful human motivation: curiosity, or fear?

96 Could You Release a "Gunpowder Microbe"?

You are a scientist using gene splicing to create new forms of virus. Unexpectedly, you develop a microbe that feeds off the minerals that constitute gunpowder. The virus is very tough—it can lie dormant for years, but still survive. It is also extremely small, so it can penetrate almost any tiny crevice.

If you release this virus from your laboratory it will multiply and spread freely, seeking out all supplies of explosives and transforming them into inert chemicals. Bullets, grenades, and bombs will be rendered useless all over the world. Even nuclear weapons will be disabled, because they require conventional explosives to initiate the nuclear reaction.

Imagine a world where there are no explosive weapons of any kind.

Will it be more or less violent than the world today?

Do you believe that nuclear weapons have maintained a balance of power, preventing world wars?

If this country loses its firepower, will another country be more likely to invade, using primitive weapons?

97 Would You Help to Develop an Addictive Entertainment System?

You are employed by a consumer-electronics company that has just developed a new kind of television. It projects images directly into the viewer's eyes, creating total realism. Everyone who sees the system wants one. When people use it, they enter a trance state and have great difficulty switching the unit off.

This gadget will be more addictive than television, just as television was more addictive than radio. Kids, especially, will find it irresistible.

How will this ultimate entertainment device change the world? Will it have educational applications? Which will people use it for: self-improvement, or cheap thrills?

Would your conscience bother you if you were the engineer who invented this device?

Would it bother you if you were merely a secretary, sending out press releases?

Do you feel responsible for the research and development that your corporation chooses to do, regardless of your job title?

Do you think scientists should take responsibility for the ways their research is applied?

98 How Should We Share Water Supplies?

In many Western states, the level of groundwater has been dropping steadily during the past few decades. If water consumption is not reduced, supplies will eventually start to dry up.

Can we deal with the problem simply by getting people to use less water for watering their lawns and washing their cars?

Who consumes more water: individuals or businesses?

Do farmers have a higher priority than other businesses when we decide who has the most need for water? Does that mean we should ration some businesses, but not farmers? And in that case, will we have to pay compensation to businesses that experience hardship as a result of the new policy?

If a resource is in short supply, how can we encourage people to use less of it, without introducing rationing?

Would there be an outcry if water became an expensive commodity?

If government made us pay a high price for water, could that money be recycled back to the community in some way?

Who really owns and cares for the groundwater that lies under the land?

99 Will You Tell Your Problems to a Robot Therapist?

Computer programs can already imitate some of the responses of a psychiatrist talking to a patient. In the future, artificial intelligence will produce a much more accurate simulation.

You visit a psychiatric clinic where you describe your problems into a microphone and receive answers over a loudspeaker. The voice sounds real, and its concern for your problems seems genuine.

How can you tell for sure whether you are talking to a person or a very sophisticated machine?

If you ask about matters of human interest that a computer wouldn't know about, and the voice refuses to talk about these topics, does this prove anything?

If you honestly believe you are talking with a person, but afterward you learn you were dealing with a machine, does this mean the machine was just as intelligent and helpful as a person would have been?

Why do you think it's easier to make a computer play chess than imitate everyday conversation?

Which requires more intelligence: playing chess with a grandmaster, or talking about someone's personal problems?

Does this mean that computers are now intelligent?

100 Will You Buy an Electronic Book?

Imagine a hand-held video display, no larger than a paperback book. The text is crisp and easy to read. Built-in batteries last one hundred hours, and you can use the gadget anywhere.

You can buy plug-in text modules to display words on the screen. Each module contains as much text as a long novel. Will you use this kind of "electronic book"?

What advantages does it have over an old-fashioned book printed on paper?

Do electronic gadgets always tend to replace old-fashioned ways of doing things?

Have the sales of pens and paper diminished as a result of the introduction of word processors? Do many people use computers to balance their checkbooks and retrieve recipes?

Can the text module for an electronic book ever be as cheap as a paperback?

You can flip through the pages of a book to find what you want. Will this be possible in an electronic book?

If a gadget is more expensive to buy and less convenient to use, will people buy it anyway, just because it's a new high-tech toy?

Will they still use it after the novelty wears off?